BUM LAMBS

by Grace Larson

Printed and bound in the United States of America
First printing • ISBN # 978-0-9908913-3-8

TO ORDER ADDITIONAL COPIES OF:

BUM LAMBS

by Grace Larson

VISIT

www.montanagracelarson.com

SCOTT COMPANY PUBLISHING
P.O. Box 9707
Kalispell, MT 59904
Toll Free: 1-800-628-0212
Fax: 1-406-756-0098

I dedicate this book to Blindy and Susie, and even the lambs that chased me. I also dedicate this book to all the Bum Lambs in the world and the boys and girls who care for them.

My Grandpa let me help him with his sheep so I learned from him, and I am grateful that he had a lot of love and patience as he taught me about sheep and Bum Lambs.

When my Grandpa was a boy in Romania he had to chase a bear away from his sheep with a stick. He loved sheep and when he came to America he had his own sheep ranch, and I was lucky enough to be his granddaughter.

Have Fun and enjoy ...... Bum Lambs...

Grace's grandfather had thousands of sheep. When the baby lambs were born some of the mother sheep (ewes) would not let them drink milk from their teats. If these little lambs didn't have milk right away they would die.

So Grace's grandfather gave Grace these little lambs. They were called Bums because they had no mothers. Every day Grace would milk the family milk cow and save the milk for her Bum Lambs. She made a bottle by using a large pop bottle and fitting a special nipple onto the end. The Bum Lambs could drink their milk by sucking on the nipple.

Grace's lambs grew very fast and they became demanding. "I want my milk now", and if Grace wasn't quick enough they would butt her with their heads, or jump up on her. Grace wasn't even 5 feet tall so they could make her lose her balance and fall backwards.

She had to be very careful when her lambs went out to eat grass and alfalfa. They could only be out for an hour or they would bloat and die. Their stomachs would swell and they couldn't breathe. One day she was late bringing her lambs into the barn. Blindy was her favorite lamb because he was chubby and also blind. Blindy had bloated and fell over dead. Grace cried and cried as she dug a grave for Blindy. She knew she had to be a lot more careful with her lambs.

Susie was a pretty lamb. She was smaller than the other lambs and she had a very pretty face. Then there were the buck lambs that grew beautiful horns. They were Rambouillet and mature.

Rambouillet rams can weigh between 250 and 300 pounds. Grace's rams were not full grown but they had the curled horns , and they liked to chase her. They knew she was afraid so they thought it great fun. One time Grace ran so fast she ran right out of her shoes. Ouch, the rocks sure hurt her bare feet.

Somehow she had to find a way to get to the barn. How could she? Maybe she could throw a rock at the rams. Maybe she could sneak to the barn so they wouldn't see her. Grace threw the rock but it missed the ram and she was knocked down. She tried sneaking to the barn but the ram saw her and knocked her down again.

Grace had been too embarrassed to tell her father she was scared of the Bum Lambs. Finally she had to tell her father what had been happening. Her father told her to get a big stick and hit the ram on the nose. Grace wasn't sure she would be fast enough but she would try.

The next time she started for the barn the ram saw her, lowered his head, and came after Grace as fast as he could go. Grace was so scared she could hardly lift the stick, and he was getting closer. "I have to do this", Grace said to herself. I cannot be knocked down again. The ram was coming so fast but Grace was finally brave enough to hit him on the nose. He stopped and stared at Grace then he turned and walked away.

Grace wasn't scared anymore but next year she was going to ask her Grandpa for girl lambs (ewes) because they wouldn't have horns, and they would not chase her.

TO ORDER ADDITIONAL COPIES OF:

BUM LAMBS

by Grace Larson

VISIT
www.montanagracelarson.com

Made in United States
Orlando, FL
04 November 2026

90208506R00017